Warren Billy Smith

UFO Investigator
or
Hoaxster?

Editor@Middle-Coast-Publishing.com

MIDDLE COAST PUBLISHING

Middle-Coast-Publishing.com

Biography

Warren Billy Smith, (Iowa, 1931–2003) was a prolific American author who wrote books on cryptozoology and the hollow earth theory.

Warren Billy Smith posing on the banks of the Mississippi River, Clinton, Iowa.

Life and career

Smith was born in West Virginia, and lived in Iowa starting in 1953, living first in Davenport, then Durant, and from 1960 in Clinton, Iowa until his death on May 9, 2003. He published many newspaper reviews of automobiles, newspaper and magazine articles on boating and other subjects, and from 1965 until near his death, a huge number (over fifty) paranormal books on topics such as Bigfoot, ancient astronauts, Atlantis and lost cities.

He is most well known for his books on the Hollow earth, and his book on Bigfoot called Strange Abominable Snowmen. His book **This Hollow Earth** was a popular book on the hollow earth theory which discussed evidence from ancient myths and legends such as the Buddhist city of Agharta to the ideas of a physical hell found in religious texts to conclude they were based on actual places inside the earth. He also discussed in the book that there are tunnels in specific locations such as mountains in South America which lead into the interior of the earth.

He published a number of books under the pseudonym of Eric Norman and he co-wrote various books with the paranormal writer Brad Steiger.

Warren Billy Smith
UFO Investigator or Hoaxster?

While our society may revere writers in general and UFO writers in particular, it's important to remember they are only human, of flesh and blood and skin and bones. As a species, freelance writers struggle in a literary ghetto. Most do not share the status or the income levels of best-selling novelists like Stephen King or Tom Clancy. Instead, writers struggle at the job of living, enduring long hours at the keyboard. At the end of the day publishers pay us a pittance for our blood. Think dozens and dozens of dollars!

Writers put up with bill collectors, shabby clothes and a rusted out, oil-burning automobiles because being a writer is a calling, like being a priest or a politician. We're helpless to do anything else. Noted UFO author Warren Billy Smith said more than once that if he had invested half of the effort he put into his writing career into a pizza franchise, he would be a millionaire.

In his long career Warren published dozens of novels, non-fiction books and magazine articles. In the UFO field, he was best known for his reporting on the Shirmer, Del Rio, Texas, and Stoughton, Wisconsin cases. The problem is that his research has always been suspect. Too many times no one has ever been able to locate Smith's conveniently transient witnesses.

Herbert Schirmer

Basically the case involved 22 year old Nebraska police officer Herbert Schirmer (born 1945), who claimed to have been taken aboard a UFO in 1967 by humanoid beings with a slightly reptilian appearance, wearing a winged serpent emblem on the left side of their chests. '

Schirmer contacted Smith after the Condon Committee sort of abandoned the case. Smith included a long chapter about the alleged abduction in his book Gods, Demons and UFOs,

Skeptics consider Schirmer's claims to be a hoax

But here today, in this article, the whole truth will be revealed for the first time. On more than one occasion Warren Billy Smith fabricated entire UFO incidents from beginning to end. I know this because Warren Billy Smith told me so on more than one occasion over the course of our 25-year friendship. This gives rise to the salient question: Why on earth would he fabricate? The answer is as simple as the sea is salt. Necessity was the mother of invention.

As he told the story, one wintry morning, Warren had a UFO book contract to finish and he was one chapter short. The deadline loomed large. He needed the back half of the book advance in order to pay bills and to buy Christmas presents for his four kids. In order to fulfill the terms of the contract, the manuscript absolutely, positively had to be on the editor's desk in New York by week's end. There was no time for further research. So Smith did what he had to: He knowingly committed falsehoods to paper. Using his well-honed fiction writing skills, Smith created two eyewitnesses to a supposed UFO event in Missouri.

As I mentioned earlier, Warren related this and other stories of deception more than once over coffee in his hometown of Clinton, Iowa. He liked to hold court at the Village Inn restaurant with its bottomless cups of coffee. The truth of the deception was corroborated by Glenn McWane, who worked as a researcher for Smith and a number of other UFO authors in the 60s and 70s. Glenn confided in me that Warren had admitted to him the deception.

That was about the time and the reason why Glenn ended his professional relationship with Warren.

Coincidentally, it was also in Clinton, Iowa that the budding UFO author Brad Steiger met Warren Smith. Back in the 60s Brad Steiger was a high school teacher who wanted to become a writer. To hear Warren Smith tell the story, he showed Steiger how to write and sell his work.

Truth be told, it was Smith who wanted to become a writer and Steiger who helped him get started.

For the record, while it's true Warren and Brad Steiger collaborated on a number of books, Brad's research and writing is now and has always been renowned for its accuracy and truth.

Allegedly the Brad Steiger Smith friendship of many years came to an end when Smith insisted that Steiger collaborate with him on creating the secret diaries of Martin Luther King, Jr. As the story goes, Smith became so angry when Steiger refused, that he delivered an ultimatum, that if I didn't do this with him, their friendship was over. Without another word, Steiger got up and left the restaurant where they were meeting and never really spoke to Smith again.

A number of people over the years have pessimistically decreed that Warren Smith was not in fact a real person, but was instead a pen name Steiger had adopted. This is false.

Understanding exactly why Warren fabricated UFO testimony comes easier when you delve deeply into his background. Warren was a child of the Great Depression. His stepfather was a veritable conman from the hills of West

Virginia who shamelessly used Warren to lend legitimacy to his scams. One favorite ruse involved parking their old Ford motorcar within sight of a farmhouse.

Methodically the stepfather stepped out of the car, took off his suit jacket, neatly folded it in half and laid it across the front seat of the car. He looked like a dandy, a man with money. Dutifully father Smith jacked up the car and changed the tire, pretending it was flat.

Finished with part one of the ruse, he and little boy Warren would casually stroll up to the farmhouse, knock on the door and ask whoever answered for a cool drink of well water. The charming, well-dressed stepfather made polite conversion for a moment or two, rattling on about the weather, the depression, and the flat tire. Then, like TV police detective Colombo, the stepfather would turn to walk away, stop in mid stride and do an about face.

"By the way," he would tell the farmer. "Took off my signet ring before I changed the tire. Stuck it in my pocket. Musta fallen outta my pocket." He sighed at the imagined loss. "Family heirloom. Find it, I'd pay a $100 reward." With that the stepfather would scribble a phony telephone number and address on a scrap of paper and hand it to the farmer.

Some weeks would pass before the partner in crime would show up at the farmer's door.

Unshaven and ragged, looking like a drifter, the partner would also ask for a glass of water. He too made small talk. In the midst of pleasant conversation he'd pull out a cheap paste ring and show it to the farmer. "Found it down there by the side of the road. Figure it's worth anything?"

Greed is the essential element in any con. How much money the ersatz bum got for the cheap ring was only limited by how much money the farmer had in his cookie jar. Sometimes it was five dollars, sometimes six. In the depression, when hamburgers cost a quarter, that was a lot of money.

Warren learned at a young age how to get money for nothing.

Another piece of puzzle relates to Warren Smith's early writing career and his association with author James Jones. After his return from the fighting in the Pacific islands during the Second World War, James Jones met a married woman named Lowney Handy. Long story short, she became his writing teacher and his lover (her husband approved of the liaison). With the blockbuster success of the novel From Here to Eternity the two of them founded the Handy Colony for struggling writers.

The way Handy saw it, "There is no more

than a hair's breadth between the artist and the criminal. The artist graduates out of the criminal class and looks into his heart and writes, or else he watches those around him with a cold clinical eye and writes about himself as he sees them." Over its 20 years, the Handy Colony attracted some 70 drifters, rebels and struggling writers. One of them was Warren Smith.

Handy forbade alcohol and rich food, and prescribed enemas for writer's block. Once a month she'd haul her troop of students across the border to a Terre Haute, Indiana brothel.

One particular technique Lowney taught was copying. Copying meant beginning each and every day by copying ten or twenty pages, verbatim, from one author or another. Who it was didn't really matter. At the end of the exercise, the pages were unceremoniously tossed into the wastebasket. Obviously this wasn't plagiarism. Instead, the exercise was intended to burn the principles of grammar, syntax and metaphor into a beginning writer's subconscious mind. Copying also breeds discipline. You write every day, ten or twenty pages. Period. No matter what. Warren Smith learned his lessons well.

Once he became a selling author, Warren Smith created a large body of original work. Non fiction credits included books on UFOs, the hollow earth, clairvoyance, the abominable snowmen, Edgar Cayce and satanic worship. His articles appeared in True, Argosy, Fate and dozens of other magazines.

He also authored dozens of historical romance and western novels, writing under his own name and a dizzying array of pseudonyms such as: Eric Norman, Robert E. Smith, Joanna Warren, Norma Warren, David Norman, Barbara O'Brien and Barbara Brooks. It should be noted his wife's first name was Joanne and his daughter's was Barbara.

This writing under a pseudonym is common practice among prolific writers. Max Brand, for example, had as many as three and four of his Western novels competing on the New York Times bestseller list.

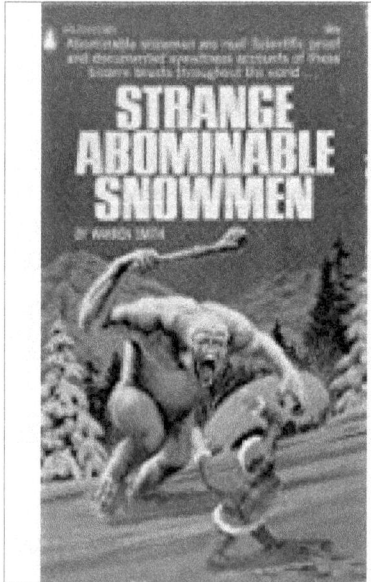

So he began publishing under the Frederick Faust nom de plume. Actually, his given name was Frederick Schiller Faust and his pen name was Max Brand. Faust wrote under 15 pseudonyms in all, including: Owen Baxter, Evan Evans, George Evans, David Manning, John Frederick, Peter Morland, George Challis, Peter Ward and Frederick Frost. You will probably want to know Brad Steiger's pen name was Eugene Olson.

Often Warren Billy Smith copied-slash-rewrote other author's works and sold it as his own. He used to brag over coffee how he rewrote novels by Stewart Edward White (circa 1900). If you didn't know the source of the original material, then the massaged work and the original appeared to be written by entirely different authors.

Interesting to note was what happened when Smith was writing Historical Romance novels for Zebra books. This genre of books were notorious for their provocative covers featuring a man and woman in a romantic embrace amidst an exotic locale. Warren Smith suggested to the art director that they airbrush the images using subliminal seduction. In this case that meant hiding breasts and male genitalia in the folds of clothing. Suffice it to say, his books sold well.

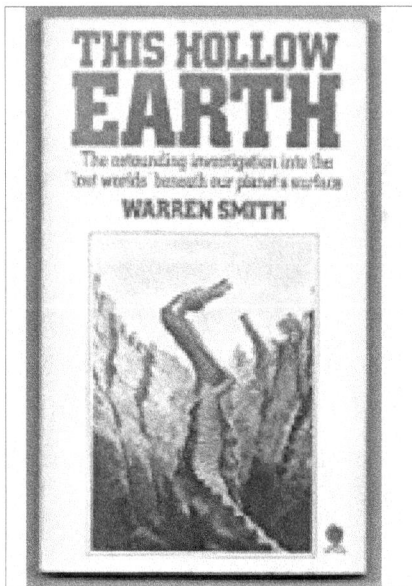

Don't misunderstand. There's no question Warren Billy Smith was a talented author. There's also no question that in at least some cases his non-fiction books and magazine articles were in reality fiction.

That is to say, if you can believe the autobiographical accounts of a man whose father was a con artist. In his heart of hearts, Warren took great pride in being the son of a con man.

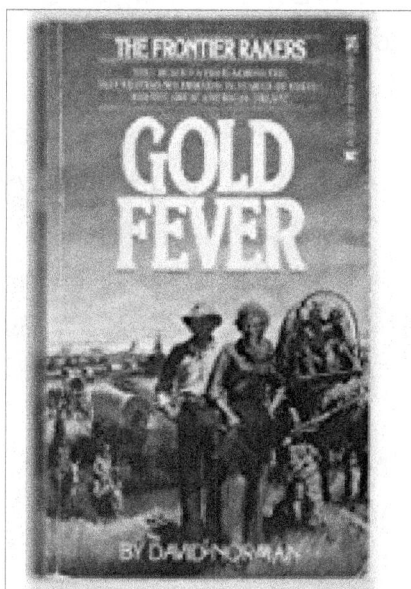

Warren Billy Smith passed away on May 9, 2003. He was 72 years old and had been in ill health for some years. He was never a step away from his oxygen bottle, having never been able to kick the cigarette habit that ultimately killed him.

Without a doubt there will be some readers who are angry, and wondering why a supposed long time friend would write this expose. The answer is that Warren would be amused by it all. Knowing him and his personality as well as I do, my guess is he's laughing at the UFO scams he successfully pulled off nearly 30 years ago. Also, the man who published more than a million words loved what he did for a living, deep down to his bones. Even if he never did invest in Pizza Hut.

No matter whether he was a UFO hoaxster or investigator, Warren Billy Smith will be missed.

Bibliography of Smith's Books

Writing under the pseudonym Johnny Shearer
Sodom, U.S.A. Brandon House 1965
The Male Hustler Brandon House 1965

Writing under the pseudonym Paul Warren
The Sensual Male Pinnacle 1971

Writing under the pseudonym Gabrial Green
Let's Face Facts About Flying Saucers. Popular Library 1968

Writing under Warren Smith:
Finder's Keepers. Belmont Books, 1965
Strange Women of the Occult. Popular Library, 1968
Strange & Miraculous Cures. Ace Books, 1968
Strange Powers of the Mind. Ace Books, 1968
Into the Strange. Popular Library, 1968
Abominable Snowmen Award 1969
Strange Murderers & Madmen Popular Library 1969
Strange ESP. Popular Library, 1969
Strange Abominable Snowmen. Popular Library, 1970
Strange Hexes. Popular Library, 1969
Talking to the Spirits: 10 World Famous Psychic Reveal Their Occult Secrets. Pinnacle, 1971
Talking to the Spirits Pinnacle 1971

Bitter Harvest Delton Press 1971
Satan's Assassins. Magnum, 1971 (with Brad Steiger)
The Strange Ones. Popular Library, 1972
Chains of Fear Delton Press 1972
Predictions for 1973 Award 1972
Predictions for 1974 Award 1973
Triangle of the Lost. Zebra, 1975
Myth and Mystery of Atlantis. Zebra, 1975
Predictions for 1975 Award 1974
Predictions for 1976 Award 1975
Predictions for 1977 Award 1976.
Secret Forces of the Pyramids. Zebra, 1976
The Hidden Secrets of the Hollow Earth. Zebra, 1976
Lost Cities of the Ancients--Unearthed! Zebra, 1976
Secrets of the Loch Ness Monster Zebra, 1976
Authentic Directory of CB Award 1976 UFO Trek
Zebra 1976
UFO Trek. Sphere, 1977
Ancient Mysteries of the Mexican and Mayan
Pyramids. Zebra, 1977
The Book of Encounters. Zebra, 1977
The Secret Origins of Bigfoot. Zebra, 1977
This Hollow Earth, Sphere, 1977
Dreams of Darkness, Dreams of Light in Dragon
Lance Tales, Volume 1: The Magic of Krynn . TSR,
Inc., 1987

Writing under the pseudonym Eric Norman:
The Under People. Lancer Books, 1969
Buried Treasure Guide Award 1970
Gods, Demons & UFOs. Lancer Books, 1970

Gods and Demons and Space Chariots. Lancer 1970
Beyond the Strange. Popular Library, 1972
Gods and Devils from Outer Space. Lancer books, 1973
This Hollow Earth - Lancer 1972 in 1997 Japanese edition

Writing under the pseudonym Robert E. Smith:
Doc Anderson: The Man Who Sees Tomorrow. Paperback Library, 1970
Doc Anderson: We Live Many Lives. Paperback Library, 1971
Doc Anderson: The Healing Faith. Paperback Library, 1972

Shared byline with Brad Steiger:
The Menace of Pep Pills. Merit Books 1965 (Warren Smith & Eugene Olson)
What the Seers Predict 1971 Lancer Books 1970
Satan's Assassins Lance 1971
Predictions for 1972 Lancer 1971

Writing under the pseudonym: Barbara O'Brien
Martinis, Manhattans or ME? Zebra 1974

Writing under the pseudonym: Barbara Brooks
High Society Pinnacle 1974

Writing under the pseudonym: David Norman
The Frontier Takers Zebra 1979
The Frontier Takers #2: The Forty Liners Zebra 1979

Frontier Takers #3: Gold Fever Zebra, 1980
Frontier Takers #4: Silver City Zebra1980
Frontier Rakers: Montana Pass Zebra 1982
Frontier Rakers: Santa Fe Dream Zebra 1983

Writing under the pseudonym: Norma Warren
Trails West Zebra 1985

Writing under the pseudonym: Jake Logan
High, Wide, and Deadly Berkeley Western 1987
Gold Fever Berkley Western 1989

Joanna Warren
The Conrad Chronicles: Belle Mead Zebra 1978
The Conrad Chronicles #2: The Dreamers Zebra 1980
The Conrad Chronicles #3: The Destined Zebra 1980